SPACE TECHNOLOGY

Monika Davies

Consultants

Michael Neufeld
Senior Curator
National Air and Space Museum

Cheryl Lane, M.Ed.
Seventh Grade Science Teacher
Chino Valley Unified School District

Michelle Wertman, M.S.Ed.
Literacy Specialist
New York City Public Schools

Publishing Credits

Rachelle Cracchiolo, M.S.Ed., *Publisher*
Emily R. Smith, M.A.Ed., *SVP of Content Development*
Véronique Bos, *VP of Creative*
Dani Neiley, *Editor*
Robin Erickson, *Senior Art Director*

Smithsonian Enterprises

Avery Naughton, *Licensing Coordinator*
Paige Towler, *Editorial Lead*
Jill Corcoran, *Senior Director, Licensed Publishing*
Brigid Ferraro, *Vice President of New Business and Licensing*
Carol LeBlanc, *President*

Image Credits: p. 2, 4–6, 7 (bottom), 8–12, 14–25, 32 NASA; p. 5 Smithsonian Institution; p. 8 Alamy; p. 21 (inset) National Portrait Gallery; p. 24 Getty Images; all other images iStock and/or Shutterstock, or in public domain

Smithsonian

TCM Teacher Created Materials
5482 Argosy Avenue
Huntington Beach, CA 92649
www.tcmpub.com
ISBN 979-8-7659-6890-1

Printed by: 51497
Printed in: China

Table of Contents

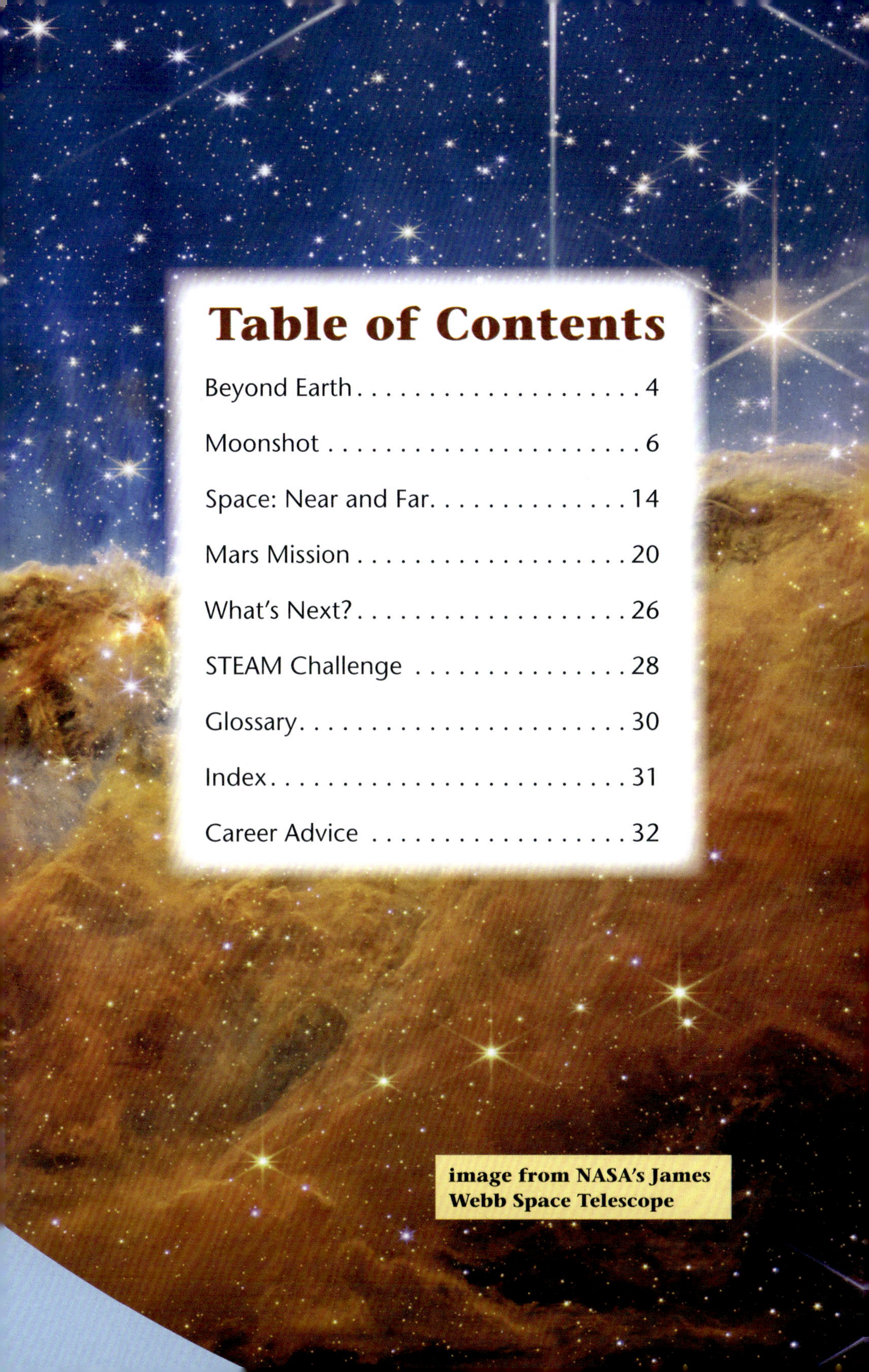

image from NASA's James Webb Space Telescope

Beyond Earth

For centuries, humans have looked to the skies and wondered what lay beyond the stars. The universe surrounding Earth has been both awe-inspiring and **elusive**. For thousands of years, humans could only scan the skies with their eyes. Then, in 1610, Galileo Galilei used a telescope to peek into the greater universe. He identified stars and discovered features on the moon.

International Space Station

Thanks to space technology, humans have started to explore space. Space technology is any technology that is used beyond Earth. This includes a mix of tools and software. Space technology consists of satellites, **rovers**, shuttles, and much more. These tools have given people answers about the universe up close and their place in it.

In 1957, the first piece of space tech began to **orbit** Earth. Since then, humans have made tracks on the moon. An entire space station was built to orbit the planet. And rovers have rolled and researched across Mars.

Without space technology, space exploration would not exist. Many discoveries would remain uncovered. In the past, space technology has guided humans to new places. In the future, space technology holds a lot of potential for exploring the universe.

Sputnik 1

Curiosity **Mars rover**

lunar rover

lunar module

Moonshot

Earth only has one moon. This bright light in the night sky has captivated humans for centuries. In 1962, the United States set a lofty goal: put a human on the moon. This led to the Apollo Program, a plan for human spaceflight. This program existed between 1960 and 1972, and it flew astronauts 11 times. Countless hours and immense efforts were invested in the program. Getting to the moon required **innovative** minds and technology.

On July 20, 1969, more than half a billion people held their breath as they watched the Apollo 11 mission on television. People worldwide watched as Neil Armstrong put the first human footprint on the moon. Viewers cheered as he said, "That's one small step for man, one giant leap for mankind."

The *Saturn V* rocket was used for several Apollo missions.

Rockets and Orbit

Gravity keeps people grounded on Earth. To go beyond the planet, humans need to overcome Earth's gravity. A spacecraft alone is not powerful enough to accomplish this. So, rockets are required to blast into space. Rockets carry spacecraft, which quickly boost up into Earth's atmosphere. Once a rocket is a certain distance away from Earth, it releases the spacecraft. The energy from the rocket gives the spacecraft speed to move in one direction. Meanwhile, Earth's gravity pulls on it in another direction. The balance between these forces keeps the spacecraft in orbit.

FUN FACT

As of 2022, you can still see the astronauts' tracks from 1969 on the moon! We know this thanks to NASA's Lunar Reconnaissance Orbiter. The orbiter took images of the moon that show the tracks are still visible. There is no air or water on the moon to erode them, so it will take an extremely long time for the tracks to vanish.

"That's one small step for man, one giant leap for mankind." —Neil Armstrong

Saturn V

For nine Apollo missions, the *Saturn V* rocket played a **pivotal** role. It served as the push that got astronauts to the moon. This towering rocket stood taller than the Statue of Liberty in New York. Built to thrust a spacecraft into orbit, it was the most powerful rocket of its time. There were three stages, or parts, to the rocket. The first stage sat at the bottom of the rocket and contained fuel and five F-1 engines. These five engines produced a tremendous amount of power!

The second stage contained five J-2 engines, while the third had a single J-2 engine. In each stage, the engines used fuel until they ran out. Then, that part of the rocket would release, and the engines that made up the next stage would start. Without this powerful three-part rocket, astronauts would have never gotten to the moon.

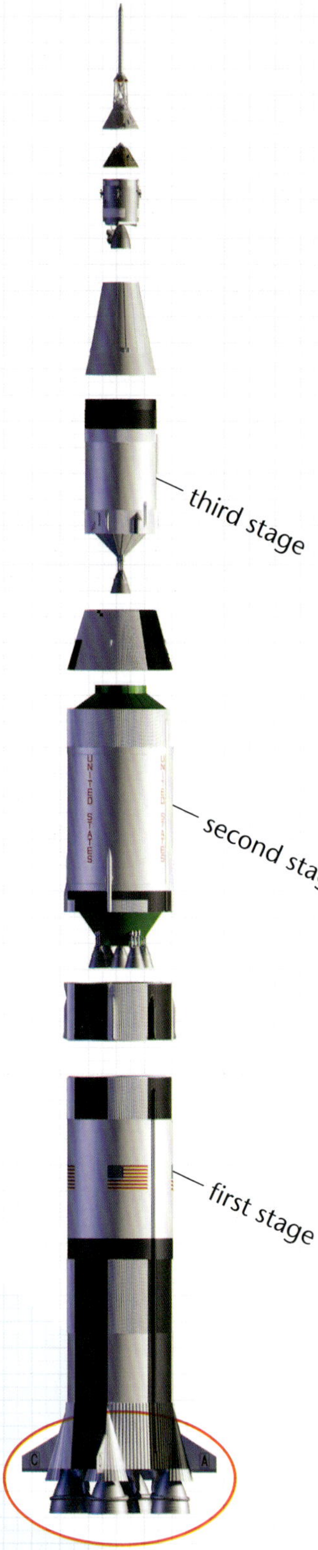

rocket engine and exhaust pipes of *Saturn V*

Modules, Modules, Modules

At the top of every rocket sat an Apollo spacecraft. While rockets helped launch astronauts into the atmosphere, the spacecraft was the technology that carried them beyond Earth.

The Apollo spacecraft was also made of three parts, or modules. First, there was a cone-shaped command module. Behind the command module sat the service module. The lunar module was then attached to the service module. Each module served a purpose and was instrumental in getting astronauts to the moon.

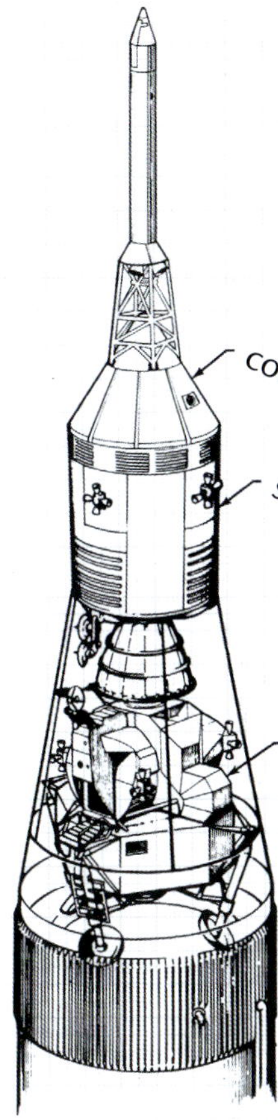

ARTS

NASA Art Program

In 1962, the head of NASA, James Webb, created a new program at NASA. It was known as the Artist's Cooperation Program. Multiple artists began to document the goal of putting a human on the moon. Some artists painted pictures of launch headquarters. Other artists painted America's first spacewalk. These paintings helped show the creativity behind space missions.

oil painting of a rocket launch

Command Module

The command module was a temporary home for three astronauts on every Apollo mission. This is where the astronauts ate, slept, and stayed in touch with people on Earth. The interior was about the size of a large car. A computer and instruments were also housed inside this module.

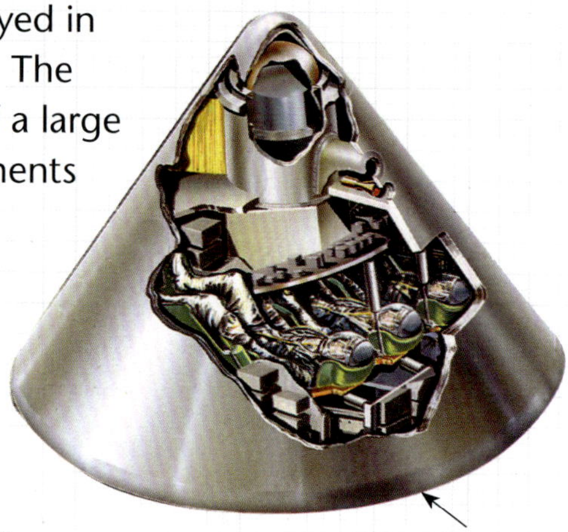

The command module was the only module used for the return trip to Earth. Since it was made for re-entry into Earth's atmosphere, it had a heat shield on its base.

Service Module

The service module was a smooth cylinder full of consumables. Consumables were anything needed to make the journey to the moon, such as oxygen, water, and fuel. This part of the spacecraft was left behind to burn up in the atmosphere.

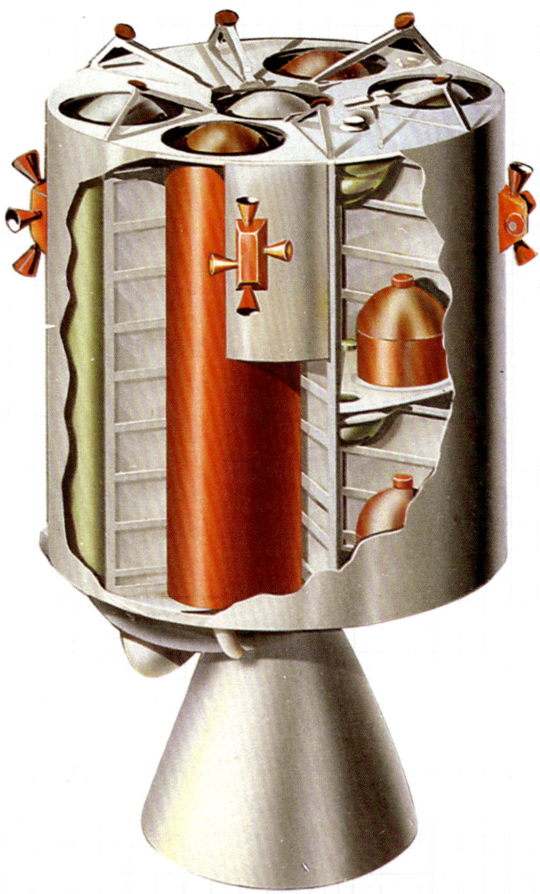

Lunar Module

The command and service modules helped astronauts reach orbit around the moon. But the lunar module enabled them to actually step foot onto the moon. This specialized module could carry only two people. It was built for service in space and on the moon. It had a lower part with landing gear, a rocket engine, and experiment equipment. The upper part returned astronauts to the command module. This part had a different rocket engine and space for the crew.

ENGINEERING

Light Lunar Landings

The Grumman Aircraft Engineering Corporation built the lunar modules that landed astronauts on the moon. This company built 15 modules, but only 8 went to the moon. Their biggest challenge was to lighten the overall weight of the module. To make the module lighter, they took out the seats for astronauts. That's right—astronauts stood while landing on the moon!

Lunar Roving Vehicle

The command and service modules were **aerodynamic**. They were built to fly through Earth's atmosphere with little **friction**. But lunar modules and roving vehicles had unique looks and purposes. They were designed for the main task of every Apollo mission—exploring the moon.

Lunar roving vehicles were a part of the final three Apollo missions. Powered with batteries, the four-wheeled vehicles bumped along the moon's **craters**. Each vehicle used a steering stick rather than a wheel for maneuvering. The vehicles could only go about 12 miles (19 kilometers) per hour. Each one was built to carry up to 1,000 pounds (454 kilograms) of astronauts and moon rocks. By the end of the Apollo missions, more than 800 pounds (363 kilograms) of lunar soil and rocks had been brought back to Earth.

Apollo 17 lunar roving vehicle

Why the Moon?

The last year humans explored the moon was 1972. And, decades later, the moon remains the only **terrain** beyond Earth with human footprints on it. But the hope is for astronauts to set foot on the moon again soon.

There are many compelling reasons for a return to the moon. Some scientists think it is an ideal location to test-drive technology that could take humans to another planet, such as Mars. Humans on the moon would be able to explore and conduct experiments. And other scientists think getting to the moon inspires people to dream beyond Earth.

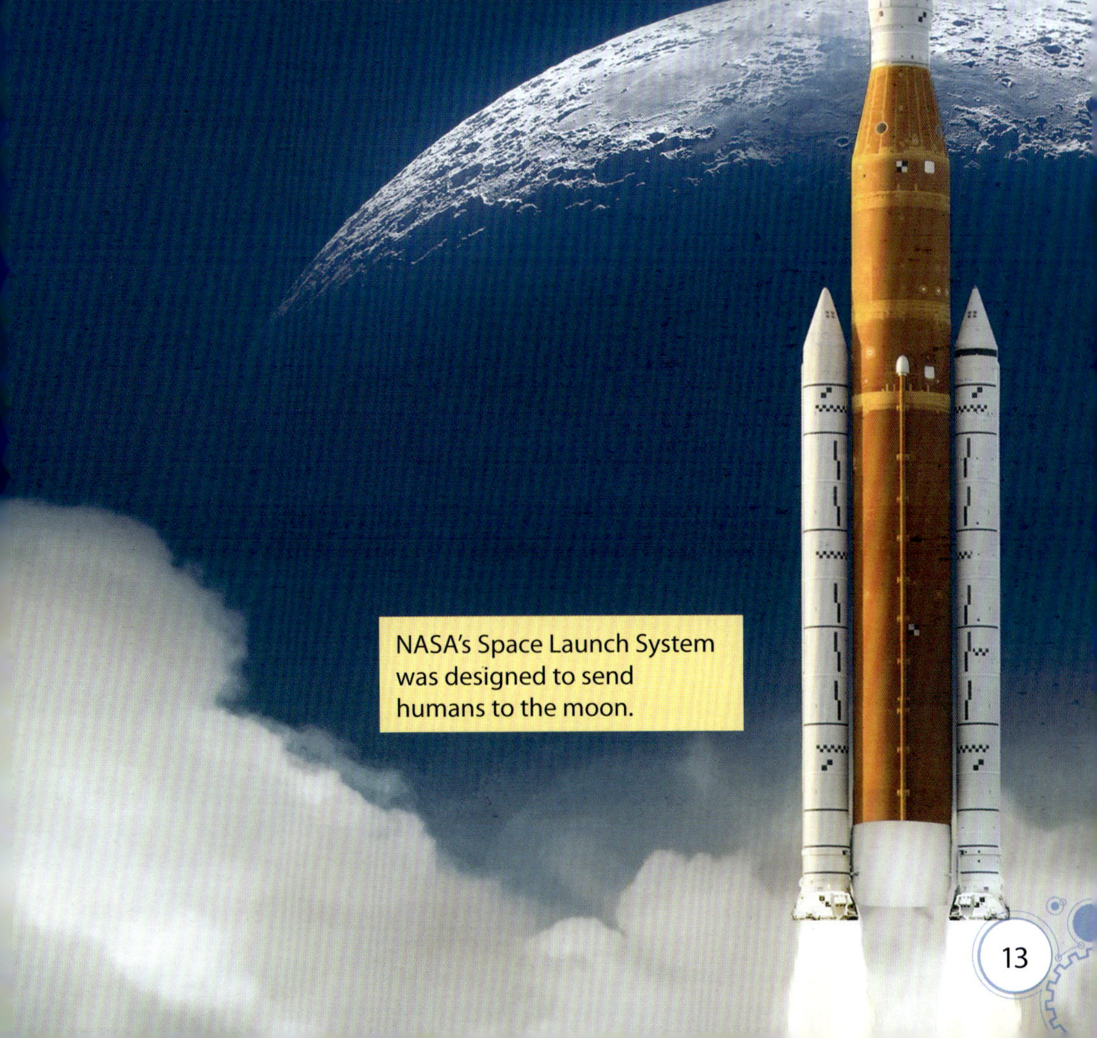

NASA's Space Launch System was designed to send humans to the moon.

Space: Near and Far

Human spaceflight is just one small part of space technology. In fact, more than 90 percent of everything done in space has little or nothing to do with human spaceflight. And since astronauts last landed on the moon, humans have remained close to Earth. Humans have not gone beyond an **altitude** of around 400 miles (650 kilometers) into space. Although a lot of space technology remains close to Earth, it still helps people explore and understand the universe.

Satellites

Artificial satellites are humanmade objects that orbit planets or moons. In fact, the first piece of technology that went into orbit was a beach-ball-sized satellite named *Sputnik 1*.

Satellites have multiple purposes. Many satellites face Earth. Some of them are used to help with communications on the planet. Other satellites collect data about Earth, such as information about our weather. Other satellites point toward the wider universe. These might monitor asteroids or photograph planets and stars.

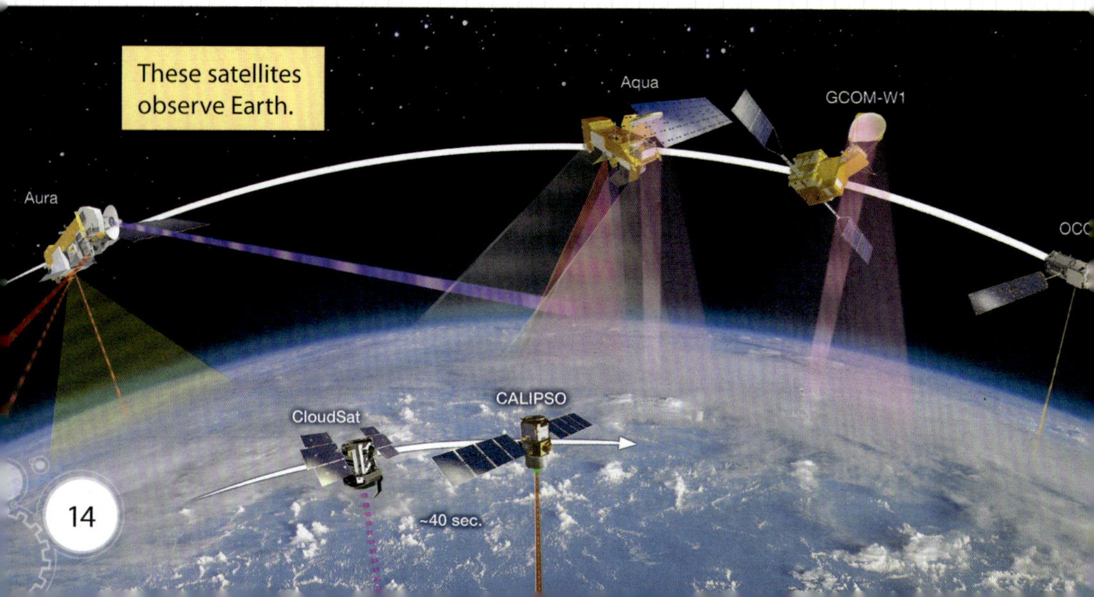

These satellites observe Earth.

There is no set blueprint for a satellite. But a satellite usually has an antenna and a power source. The antenna allows it to communicate with technology back on Earth. Satellites often use solar panels or batteries to stay powered up.

satellite image of a typhoon

FUN FACT

Satellites can be pieces of space technology that orbit Earth. But the word *satellite* can also be used to describe a moon. Moons, or natural satellites, are found across the solar system. Earth only has one moon, but some planets have more. According to astronomers, Jupiter has 95 moons!

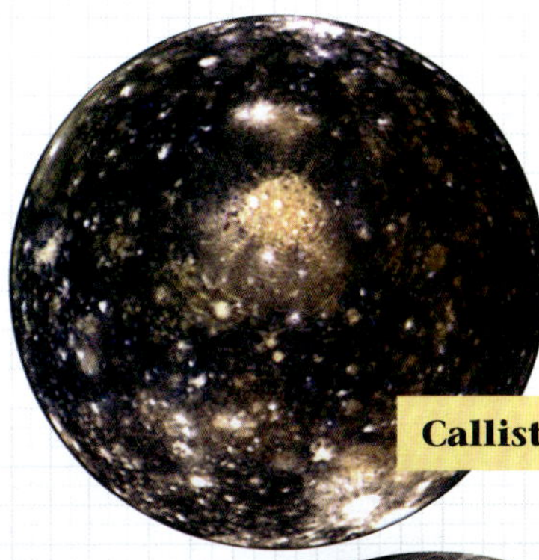

Callisto

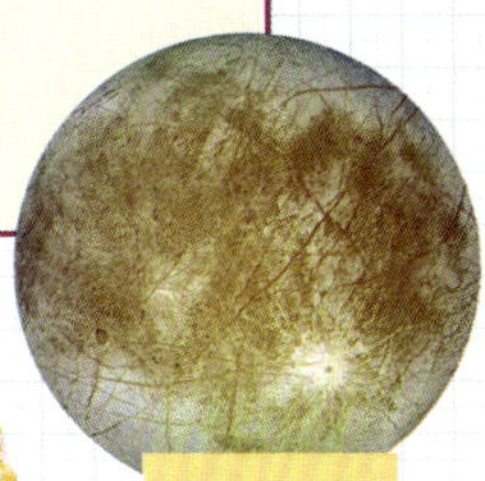

Europa

Ganymede

Io

These are the names of Jupiter's four largest moons.

Space Shuttles

From 1981 to 2011, astronauts flew into orbit on NASA space shuttles. These winged vehicles had their own rockets. Space shuttles could go to low-Earth orbit and the space station.

Every shuttle had three parts. First, there was an orbiter shaped like an airplane. It carried people, cargo, and three large rocket engines. A tank held fuel and an **oxidizer**. Two rocket boosters hugged the tank. These helped launch the shuttle.

Five space shuttles flew in space. *Discovery* was the third of its kind. It had a total of 39 missions and spent a total of a year in space.

Endeavour (left) is docked at the International Space Station.

Discovery flies over Washington, DC, on a NASA plane.

Hubble Space Telescope

Humans can see some celestial details from Earth's surface. However, the planet's atmosphere obscures our view. Scientists realized they could capture crisper images of space by placing a telescope *in* space.

In April 1990, a very large **optical telescope** was launched above Earth's atmosphere. It was named Hubble. It can detect different wavelengths of light, including light that is not visible to human eyes. Hubble takes pictures of objects in space. In its lifetime, it has made 1.5 million observations. Hubble's discoveries have helped create a 3D map of **dark matter** and spotted two undiscovered moons around Pluto. Hubble has even helped scientists learn that the universe is about 13.8 billion years old!

MATHEMATICS

Meet a NASA Mathematician

Josephine Jue was one of the first Chinese Americans to work for NASA. In 1963, she was hired as a mathematician and computer programmer. In the 1970s, she helped develop a computer programming language. It was called the *HAL/S compiler.* It was used in software for many U.S. space projects, including space shuttles.

The International Space Station

The International Space Station (ISS) is the largest example of space technology. The station is a huge laboratory that orbits Earth. Forward-thinking research is conducted here. Over 250 astronauts from 21 countries have spent time on board.

The ISS took 10 years to build. The United States and Russia launched different modules, or parts, into space. These modules were connected to make up the station. Astronauts from many countries helped build and continue to maintain the ISS.

The ISS is an intricate piece of space technology. The station itself is 356 feet (109 meters) long—almost the length of an American football field. Most of the time, the ISS houses around seven people, though it can hold more. The station uses solar panels to generate the power needed to run the station. Over 50 computers manage the different systems on the ISS. Around 1.5 million lines of computer code run on those computers. The ISS has many complex systems!

In space, people or objects appear to be weightless. This condition is known as *microgravity*. A space laboratory on board the ISS offers the chance to experiment in this unique environment. To date, astronauts have carried out over 3,000 experiments at the ISS. Some of these experiments have led to new insights. Many of them have improved life for people on Earth.

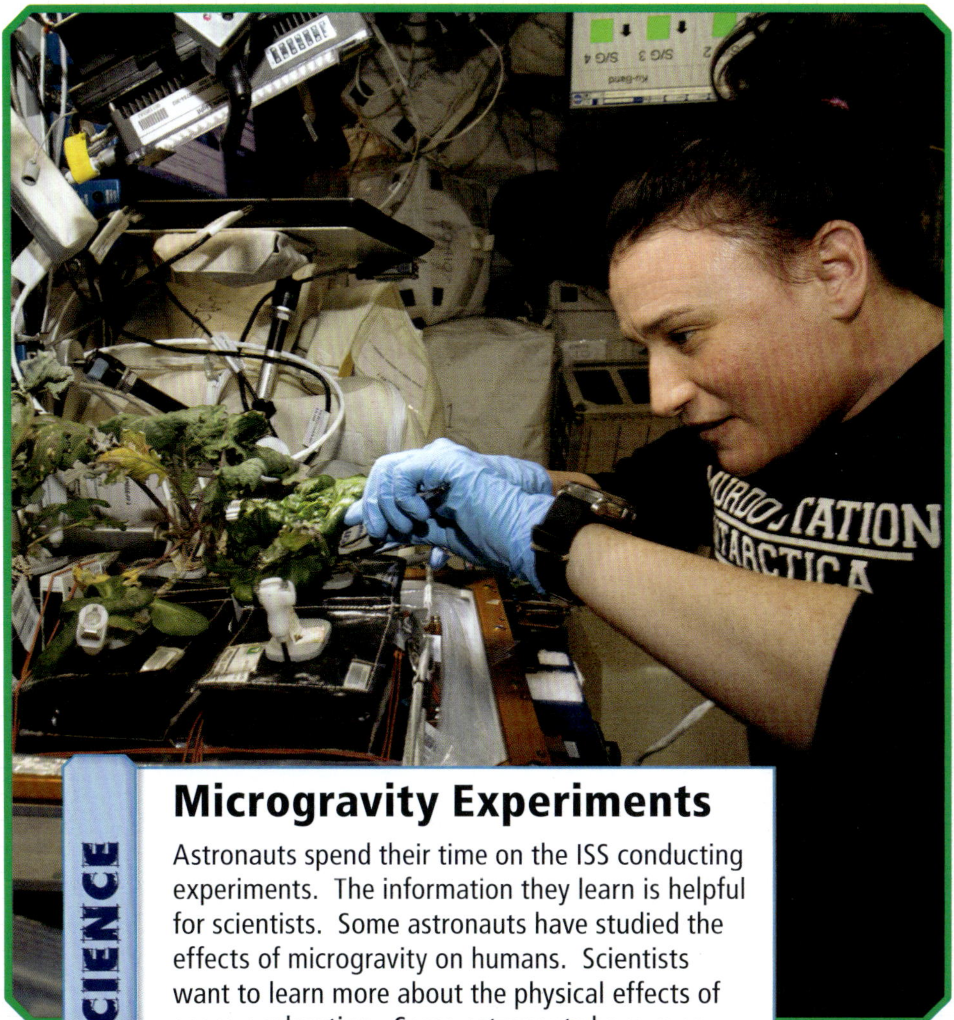

SCIENCE

Microgravity Experiments

Astronauts spend their time on the ISS conducting experiments. The information they learn is helpful for scientists. Some astronauts have studied the effects of microgravity on humans. Scientists want to learn more about the physical effects of space exploration. Some astronauts have even grown plants in space! Scientists want to study how the conditions in space affect plant growth.

Mars Mission

As humans continue to explore space, the planet Mars remains a focus. The red-hued planet shares some exciting parallels with Earth. Billions of years ago, lakes and rivers crisscrossed the landscape of Mars. The Martian atmosphere was once thick enough to sustain liquid water. This fact alone has fascinated humans.

Today, Mars is a very different planet. But it offers a window to the past—and possibilities for the future. Space technology, including robots, helps people on Earth explore their planetary neighbor.

The First Arrivals

The first piece of space tech to land on Mars was the Soviet *Mars 3* probe in 1971. Unfortunately, it failed 110 seconds after it landed. In 1976, NASA was successful with two landers. They were *Viking 1* and *Viking 2*. These landers took the first pictures from the surface of Mars. They also looked for possible signs of life on the red planet.

The surface of Mars has many features, including polar ice caps, canyons, and craters.

Meet *Sojourner*

In 1997, *Sojourner* was the first rover on Mars. This six-wheeled vehicle landed on Mars using an airbag system to cushion its arrival. It had an x-ray **spectrometer**. The first images it took were of rounded rocks on Mars. These rocks indicated there was once liquid water on the planet. Water is capable of causing rocks to become smooth and round over time.

Sojourner was about the size of a microwave oven. The rover was less sophisticated than current Mars rovers. But it showed how technology could help scientists learn more about a mysterious planet.

FUN FACT

Sojourner got its name thanks to a writing contest. Students around the world wrote essays describing important heroines throughout history. The essays had to explain why the rover should be named after the heroines. In the end, a student who wrote about Sojourner Truth won the contest. Sojourner Truth was an abolitionist who fought for civil rights. Her first name means "traveler."

Meet *Spirit* and *Opportunity*

The Mars Exploration Rovers followed after *Sojourner*. *Spirit* and *Opportunity* touched down on Mars in 2004. Both had more modern technology compared to the last Mars rover.

The two rovers shared a mission. Scientists were curious about the existence of water on Mars. They also wanted to learn more about the planet's climate. So, each rover had high-quality cameras. Different spectrometers were attached to each rover. The rovers also had tools they could use to dig into rocks. They were each about the size of a golf cart—much larger than their sibling, *Sojourner*.

Spirit* and *Opportunity

Scientists learned a lot from their findings. *Spirit* examined rocks and took pictures of them. *Opportunity* found more evidence that liquid water had been on Mars. *Opportunity* also discovered gypsum. This is a type of mineral. Gypsum can only form when liquid water is present.

Meet *Curiosity*

In 2011, the *Curiosity* rover landed on Mars. Again, this rover came with new technological advancements. *Curiosity* had **radiation** detectors. It also used **sensors** to learn more about the environment and atmosphere.

When *Curiosity* entered new terrain, it scanned the area with its cameras. It used a laser sensor to determine which rocks to look at. It then got up close and personal with these rocks using its robotic arm. *Curiosity* has made a long list of discoveries. To this day, the rover is still hard at work on Mars!

Curiosity

Meet *Perseverance* and *Ingenuity*

Perseverance (*Percy*) and *Ingenuity* (*Ginny*) are two robots. These innovative pieces of space technology are the latest robots to explore Mars. *Percy* is yet another rover, while *Ginny* is its helicopter sidekick. *Percy* drives with more speed and ease than *Curiosity*. *Percy* also has a system to collect and store rocks from Mars. *Percy's* main role is to return rock samples from the red planet to Earth. Scientists can then more deeply analyze the rocks.

Perseverance

Many scientists were unsure if flight was possible on Mars. But *Ginny* proved it could happen. *Ginny* is a small helicopter that has four legs. The solar-powered helicopter is officially the first object to fly on another planet. *Ginny* helps give *Percy* a bird's-eye view of what lies ahead. The helicopter scans areas for *Percy's* potential travel destinations. Scientists can then plan effective routes for their rover. *Ginny* also captures **aerial** shots of the Martian landscape.

Ingenuity

Mars Mysteries

Did Mars once support life? Is it a **hospitable** place for humans in the future? There are many unanswered questions about the red planet. But space technology, such as these rovers, is starting to give scientists some answers.

image from *Perseverance*

image from *Ingenuity*

TECHNOLOGY

Flying on Mars

Mars has a very thin atmosphere. Because of this, it is harder for aircraft to lift off the ground. *Ginny* the helicopter was designed to be light enough to fly on Mars. *Ginny* is made of miniaturized computers and other parts. In total, *Ginny* only weighs 4 pounds (1.8 kilograms)!

What's Next?

So much of space remains a mystery to humankind. Using space technology, scientists think they have explored about five percent of the universe. And this five percent estimate might be generous!

Space technology has evolved rapidly over the years. Less than a century ago, the first satellite was sent into space. Successful lunar landings came next. Now, an international space station orbits the planet. And rovers comb the landscape of faraway Mars.

Space technology will help people reach new horizons in the future. The number of satellites in space is likely to increase. Reusable rockets are becoming a priority. These kinds of rockets make it cheaper to launch space technology beyond Earth.

People are also building technology to help clean up messes left behind in space. This is known as orbital debris removal. For instance, many satellites are no longer in use. But they are still in orbit and might collide with other space technology. This is becoming a key issue to address.

Space continues to ignite curiosity and wonder for many people. Space technology allows us to explore beyond our planet and into an unknown universe. Only time will tell what we will discover next!

STEAM CHALLENGE

Define the Problem

NASA is planning to send more advanced rovers and other technology to Mars. Astronauts will be on board, too. The spacecraft carrying precious cargo needs to be able to slow down and land safely. Before NASA builds a full-scale spacecraft, they want to see a smaller model in action. To mimic the technology, you must design a model of a Mars lander that can safely land a marshmallow when dropped to the ground.

Constraints: You may only use the materials provided to you. There can be no lid on the cup containing the marshmallow.

Criteria: A marshmallow will represent delicate space equipment or astronauts. It will be placed in a cup to represent a cabin. When dropped from 2 feet (61 centimeters), the lander should remain upright and the marshmallow should stay in the cup.

Research and Brainstorm

How can you keep the lander from tipping in the air as it drops? How can you design parts to absorb the shock of landing on the ground?

Design and Build

Sketch two or more designs for your Mars lander. Label the parts and the materials. Choose the design you think will work best. Then, build your Mars lander. You can do a few mini drop tests as you build and make small adjustments to your design.

Test and Improve

Place a marshmallow inside the cup. Drop your Mars lander from the required height. Did the marshmallow stay inside? Did the lander stay upright? What can you add, remove, or change to make it work when dropped from a higher height? Modify your design and rebuild it as needed. Reassess how well it meets the criteria.

Reflect and Share

What other ways would you want to test your design before someone builds a real-life version and sends it to Mars? What part of this challenge are you most proud of? What problems did you encounter and how did you resolve them? What did you learn that you could apply to other challenges?

Glossary

aerial—of, relating to, or occurring in the air

aerodynamic—the quality of an object that affects how smoothly and easily it can move through the air

altitude—the height of an object above sea level

craters—large holes on the surface of a place

dark matter—a form of matter in the universe that does not absorb, reflect, or emit light and is thought to account for various gravitational effects

elusive—hard to capture

friction—the force that resists motion between objects in contact

hospitable—offering a pleasant environment where living things can live or grow easily

innovative—featuring new, original ideas

optical telescope—telescope that gathers and focuses light for viewing magnified images

orbit—travel in circles around another object

oxidizer—a substance that helps add oxygen to another substance

pivotal—incredibly important

radiation—powerful energy radiated in the form of waves or particles

rovers—vehicles used to explore the terrain of a moon or planet

sensors—devices that sense heat, light, sound, motion, etc. and respond to them in a certain way

spectrometer—an instrument used for measuring wavelengths of light spectra

terrain—a geographic area

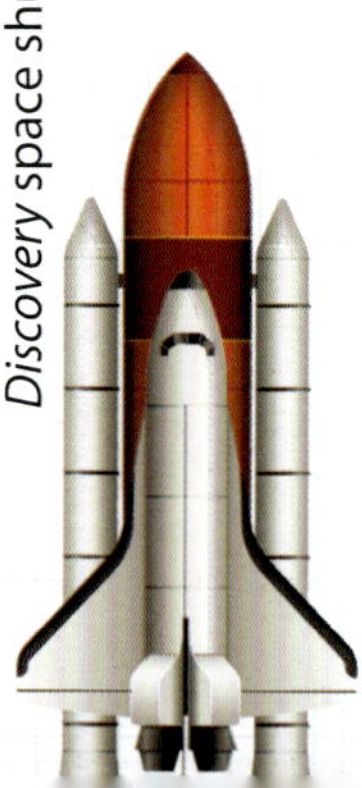

Discovery space shuttle

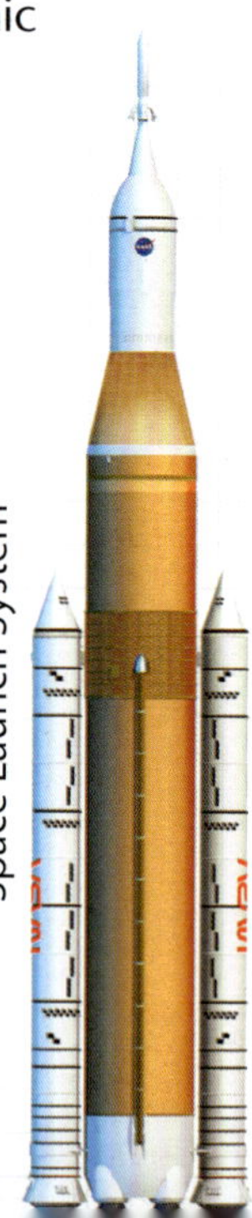

Space Launch System

Index

CAREER ADVICE

from Smithsonian

Do you dream about working on space technology?

Here are some tips to keep in mind for the future.

"Make sure to read as much as you can about math, science, and space."

– ***Michael Neufeld**, Senior Curator,*
National Air and Space Museum

"Follow your passions and really dig into whatever you choose to do. Being a good problem solver is an asset no matter what you do."

– ***Margaret Weitekamp**, Chair of Space History and Curator,*
National Air and Space Museum